THI
FOLLOWED
SCIENCE

I challenged the system for the
safety and well-being of my coworkers,
when everyone was too afraid
to say the obvious

DR. RYLEN

Published by: Book Writing Founders
www.bookwritingfounders.com

Dedication

This book is dedicated to the memory of all the fallen miners

Authors Bio

Rylen Pedneault, a 37-year-old with nearly two decades of experience in building mines, is a passionate, loving, and resilient individual. Married to Samantha Pedneault for 7 years, they have a beautiful rural home outside Sudbury, Ontario, where they share their lives with their children Layla, Oliver, and Chase. Rylen's leadership qualities have been evident since childhood, as he has always taken on roles of responsibility and embraced his patriotic nature.

Despite being known as the class clown in social settings, Rylen has also faced significant sadness in his life. However, he channels his emotions through his witty and Jim Carrey-inspired humor. Teaching is another aspect that Rylen finds fulfilling, displaying his confidence and desire to share knowledge. At times, he can be gentle and deeply emotional, particularly when reflecting on his personal journey. As a problem solver, motivator, and inventor, Rylen believes in the power of positive thinking and persistence, even in challenging circumstances.

Though mostly driven by love, Rylen acknowledges that he experiences occasional moments of negativity. He describes himself as a momma's boy, yet transforms into a fearless leader on the battlefield of life. Rylen's artistic talent shines through his ability to draw and create portraits, a skill he honed during childhood and occasionally revisits in adulthood.

Rylen Pedneault is a dedicated and multifaceted indi-vidual who combines his vast experience in the mining indus-try with leadership skills, emotional depth, and a strong belief in the importance of critical thinking and truth. His ability to navigate challenges, passion for teaching, and artistic talents make him a unique and inspiring author.

Table of Contents

Chapter 1

If I May.

When I was ten, life presented itself as an endless journey of discovery - yet opportunities to quench my burning curiosity were few and far between. The prospects could have been limited by intimidating adults or a lack of resources, but ultimately it all came down to confidence: the belief that we can find our way in this world. Thank-fully for me, there's always been something fanning the flames – 'the curiosity agent' - propelling me forward through life with one simple mantra: If I may.

Conformity has its place in society, no doubt. But I think if we rely on it too much to stifle original thought and creativity well then, humankind may as well say goodbye to advancement for good. Not only is conformity limiting for individuals like me who want to express themselves differently or push the boundaries of what's possible in the world... but it could also be our downfall when we are trying to make progress toward a brighter future down the road. As a result, we must find a balance between structure and originality if we want to continue to thrive as a species.

According to my observation, the last two decades have seen sweeping changes across a number of different areas, from political and social movements to techno-logical advances. Since the 1990s, many aspects of life have undergone drastic transformations, largely due to societal recommendations and legislative updates. For example, in the 90s, I remember riding in a box of a truck right downtown. No seatbelts are needed. I remember smoking was commonplace in many public places; nowadays, it's illegal or heavily restricted in almost every area.

With the development of technology, the last couple of decades have seen vast improvements. Not only can we process data faster and more efficiently than ever before, leading to valuable medical discoveries and scientific breakthroughs, but measures are also being taken to increase safety. Awareness around personal safety has increased with legislation such as seatbelt laws and car seats for children. It's hard for me to believe how much our lives have changed in such a short span of time - sometimes it feels like just yesterday that I was still riding my bike around, with my friends. It's amazing to consider the strides we've made both technologically and concerning personal safety since then.

One thing I've learned is our health care in Canada is one of the worst and I've worked overseas. Doctors in other countries thriving in the medical field told me our country's health care is horrible. I don't want to dwell on this idea but, for those that didn't know doctors prescribe certain drugs due to their affiliations and the commissions they gain from getting certain products sold. Doctors must advocate for our health, not just experts in a particular medication brand. It is essential for them to understand the patient's needs and suggest the best possible approach to treat a condition that does not prioritize corporate interests over the safety of individuals. Of course, this might not always be the case due to legalities, but it should still remain an effortless goal.

I would like to point out this one situation where the school boards were themselves associated with the hospitals and compelled the parents to prescribe a certain piece of medication as they could not control certain children. They needed the assistance of this specific drug to ensure the teachers could perform their job seamlessly.

Nothing ever stopped me from being successful in school. I was a decent student who managed to achieve good grades while staying on top of my passions. In addition, I tried to maintain a respectful relationship with all of my teachers. Yet, I think my curiosity is something that was looked down upon.

When it came to curiosity, no one could match my enthusiasm for asking questions. Every explanation and idea that was presented sparked new queries in me. Despite this, my parents were still convinced that I had to take medication in order to be a proper member of the class. However, this was just their effort to quiet me down and undermine my eagerness to learn. Nothing could stop me from digging deeper into an issue and finding out more information - I was determined not to let anyone hold me back.

I believe education is a noble profession, but it turns out that not all educators are living up to their ethical code - what else would explain why they're handing out drugs like Ritalin as if they are candy? It's concerning reckless behavior on the part of teachers who may think this will help students score better on tests. But at what cost, one must ask: these pills can have long-term consequences for students' health and development, which could cause real trouble for those same educators down the line with concerned parents or officials. Smarten up, folks — your school's results don't matter more than an individual student's well-being.

As a student faced with these educators, it was a disheartening feeling for me. To have my voice and opinion disregarded, simply because it was different or didn't fit the status quo in their minds - something I had done my very best to try and come to grips with. My dissent had to be restrained, no matter how much believed otherwise, as these teachers seemed to pride themselves on publicly shaming students into submission. This experience felt reminiscent of a scene from a medieval drama; more Game of Thrones than the real-life school where everyone should feel accepted, safe, and respected.

Ah, the joys of public humiliation. Like a modern-day inquisition, I was grilled like some kind of criminal for daring to speak my mind in class. All that without a shred of evidence and 'validity' on their part – how delightfully biased.

As the years went by, progressing grade after grade, I reached the ninth grade. Being a researcher, I went further with my understanding to comprehend what was this drug composed of and how it impacted me. Upon my inquiry, I found out these drugs were nothing less than mind-numbing agents to prevent children from active participation in class. It was composed of the endeavor of converting children into mindless human beings or robots that did not have any perspective of their own. This research is what motivated me to dig further and find out how certain chemicals influence the human body.

It's ironic that those entrusted to teach us the value of ambition and hard work are often some of its worst offenders. Sadly, it looks like many modern students will never fulfill their potential due to the example set by corrupt educators who preach one thing but practice another. It's really inspiring to see our respected, looked-up-to role models setting such a low moral bar for the rest of us. If there were an award for reckless disregard of morality and common decency in society, it looks like they would have no competition. But hey – what fun is living if you don't get to break all the rules?

This little research is what laid the foundation for me to further build on my knowledge and strengthen my evaluation. I was influenced to opt for a range of Math and Science classes. I took part in various science fairs to satisfy my inquisitiveness. I invented a driveway with a built-in heater, so the elderly wouldn't have to shovel snow. This project of mine received wide recognition and admiration for the concept that it dealt with. Sure, you could play it safe and only ever use 50% of your capacity in life. But why? We've got so much untapped potential that's just aching to be explored! Forget the sheep mentality - humans can achieve infinitely more if they put their minds to something and stay passionate about it. After all, the universe is made for curious people who want the best out of life - not necessarily "a lot," but whatever makes them truly fulfilled.

It looks like our traditional views of success are going the way of the dodo, as modern individuals struggle to redefine their lives with ever-evolving concepts and values. We may be marching forward into a brave new world -- but is it really one that's built for long-term stability or just an endless cycle of shortsighted fixes?

As the book progresses, you can follow alongside and learn how curiosity has played an essential role in my life. From contemplating things that I didn't understand to exploring the depths of my surroundings, it has cultivated an appreciation for complexity that continues to serve as the cornerstone of my journey.

Even today, I'm driven to discover new things, no matter how curious or ridiculous they may seem. That's why you have this book - a journey through my continuous process of learning and unlearning, which will leave you scratching your head in confusion... but that'll be half the fun!

Chapter 2

Family Trauma

As I was growing up, my childhood was anything but conventional. I was always seeking refuge outside of my own home, always eager to sleep over at a friend's house. It was during this time that I found myself being taken in and raised by a handful of my friend's parents. One family, in particular, was very passionate about boxing, and it didn't take long for me to become just as invested in the sport as they were. I joined their boxing club, and from there, something magical happened. For the first time in a long time, I found myself gaining a sense of achievement and purpose, a feeling that had been missing from my life for far too long.

I also feel I always had a sense of financial responsibility beyond my years. Maybe it was because my parents didn't have a lot of money or maybe it was just a quirk of my personality, but either way, I was always aware of the cost of things. That's why when it came time to choose a sport to play, I knew that boxing was something my parents could afford, and I was just happy to have found something that I was good at.

As a kid, I never thought that my love for boxing would actually come in handy. But it did. Whether it was at the park or on the street, I would often get challenged, but I never worried. Why? Because I knew how to defend myself. Boxing not only gave me the physical skills to fight back, but it also gave me the confidence to stand up for myself. I was never afraid to speak up when I saw something that was wrong or to confront someone who was trying to intimidate me. And that confidence has stayed with me my whole life. Thanks to boxing, I've never had to back down from a challenge. I've always been able to hold my own, both physically and mentally.

I always knew I was different from the other kids in school. While they were worrying about grades and fitting in, I was busy making my classmates laugh. I was the class clown, always cracking jokes and making silly faces. I loved being the center of attention, and I thrived on the energy I got from making people laugh. Of course, not everyone appreciated my humor. There were some teachers who just didn't get it, or maybe they didn't appreciate being called out on their lies. But that didn't stop me. I was a dogmatic individual, and I wasn't afraid to stand up against the system if I thought something was wrong. Looking back, I realize that my antics may have caused some trouble for those teachers I didn't respect. But at the time, all I cared about was making people laugh and having a good time. And in that regard, I was pretty popular.

When I was in school, we had a teacher who wouldn't let us go to the gym unless we were well-behaved. It wasn't long until the entire class was fed up with this punishment. I decided to take matters into my own hands and convinced my classmates to rebel against her. I encouraged them to protest by not answering questions, not participating, and just sitting there. And you know what? It worked. The teacher finally realized that we weren't going to put up with it anymore. Looking back on it, I realize that this was a defining moment for me - I learned that I had the power to influence others and enact change. It was a lesson that has stayed with me to this day.

My life changed in an instant, without warning or preparation. The bustle of school, friends, and laughter may have masked the turmoil brewing at home. One day, I returned home to find my parents arguing, and my dad leaving the room with a face glazed with anger. I had never seen him like that before. It was a strange time, with the feminist movement in full swing and the women's rights rhetoric gaining momentum. During this time there was a sense of female empowerment spreading, a narrative that women didn't need men in their lives. The beginning of the destruction of the nuclear family. The message around that time painted a vivid picture of the female experience, one that would leave a lasting impact. I think my mom just got sucked into this without any valid reason. I could see that my dad was trying to save the relationship but my mom was just over him.

In the beginning, my parents tried to fix things between them, but it was clear that it wasn't meant to be. As a result, my dad became the primary caregiver for all of us kids. It wasn't easy for him, but he took on the responsibility without hesitation. Meanwhile, my mom was battling depression and seeking help from the province and pharmaceutical companies. Unfortunately, her health suffered even more because of the experiments she participated in, but despite her struggles, my mom remained a presence in our lives.

Being in a blended family, I always knew that my two older sisters weren't my dad's biological daughters. They were his step-kids, but that didn't matter to any of us. To me, they were just my sisters, and I loved them just as much as I would have if we shared the same blood. When my mom had a mental breakdown and my parents divorced, it was a tough time for all of us. But my dad never hesitated to step up and take care of us, even taking one of my sisters with him. I'll always be grateful for his love and support, and for the way he made sure that we never felt like we were anything but a family.

In my younger years, my brother was always the talk of our city. My dad had him on skates before he could walk and eventually, people would often compare him to the next Wayne Gretzky – the greatest Canadian hockey player of all time. Everyone had high expectations for him, and it seemed like his future was predetermined. As for me, I struggled in school during that time, and the medication the doctor forced us to take only added up to my confusion. I often found myself debating with the doctors and trying to prove my point to my parents. However, things took a turn for the better when I stopped taking the medication. I suddenly found myself performing better in school, and it was as if a fog had lifted, and I could finally see things clearly for the first time.

During my early high school years, I began to see drugs popping up all around me. It started with small things like kids sharing Ritalin and other prescription drugs. But before I knew it, it seemed like everyone was selling and using drugs of all kinds. I remember constantly warning others to stay away from drugs and alcohol. Yet, amidst all of this, alcohol was sold and consumed freely, as if it were just a harmless beverage. It was frustrating to watch as kids bought and sold pills like they were trading cards. And yet, as wild as it all was, going to school and hanging out with my buddies still felt just average. It's scary to think how normalized drug use had become in our lives. My drug of choice was weed. It relaxed me and cleared my mind. It helped me with a lot of my day to day struggles. It became my choice of medication.

Even though I didn't like to drink alcohol, I still went to parties with my friends and quickly realized that being challenged at parties was going to become a regular occurrence. Despite not being the biggest guy around, I was always the first one to be taunted by the drunk guys who thought I was an easy target. "Hey, I bet I could take you down with just one punch!" they would say as if it was all just a joke to them. However, these people never seemed to learn that it never worked out well for them because I was sober and had a clear mind. As a result, my tough guy reputation only grew stronger, and soon enough, everybody knew to think twice before challenging me to a fight. Interestingly enough, this reputation was not due to any real desire to hurt anyone. It just wasn't my thing.

Looking back on when I bought my first home, I was just 19 years old at the time, and while most people my age were spending their hard-earned cash on fleeting pleasures, I had bigger ambitions. Of course, I couldn't have done it alone. I had a lot of help along the way, especially from the father of my girlfriend at the time. I'll always remember how much that meant to me. At the time sharing my home with my best friend, Matt made the experience all the more special. He helped me make the mortgage payments by paying rent, and we both learned a lot about adulting in the process.

When I first moved out of my parent's house, I was terrified. Living on my own seemed like such a daunting task, and the thought of being alone was enough to make me want to crawl back into my childhood bed. Enter Matt, my knight in shining armor and best friend since high school. He agreed to move in with me and help ease the transition, and I couldn't be more grateful. Having someone paying rent was a plus, but more than that, having a familiar face around the house made all the difference. We may not have always agreed on how to decorate, but having a good friend to share this new chapter with made it feel less like a scary venture and more like an adventure.

While I was on my new home adventure, my brother was being scouted by a hockey team from the States. It was clear that he had a natural talent and a passion for the sport. However, his career and dreams came crashing down after he made a single mistake at a party. He was persuaded to try Oxycodone with some friends, and this marked the beginning of his addiction. Drug dealers were giving kids oxies for free to get them hooked. Despite being a professional athlete who was checked regularly, he couldn't resist the pull of the drug. It soon became evident that he had a problem and it was traced in his bloodstream, leading to the end of his hockey career. Sadly, he stopped attending practices and was dismissed from the team. It was a painful lesson for our family, but we learned just how easily one wrong decision could ruin everything.

The depths of addiction can lead one down a dark and sordid path. Sadly, this was the case for him. Once an upstanding member of society, his drug addiction had turned him into a filthy animal. The once well-groomed man now resembled something akin to a wild creature, with overgrown hair, matted clothes, and a disposition to match. It was clear that he had become trapped in a cycle of addiction, unable to escape the clutches of his demons. As we watched from a distance, our hearts went out to him, hoping that one day he would find the strength to overcome his new dependency.

It was a difficult time for our family, but we were determined to help my brother get his life back on track. My dad took him in and we all rallied around him, offering support and encouragement. However, it seemed like trouble was always lurking around the corner. Sneaky individuals would show up at our house, and we soon discovered that my brother had been selling off our belongings to feed his addiction. It was a painful realization, but we knew that we couldn't just sit back and let it continue. There had to be a confrontation, no matter how uncomfortable or difficult it might be. We were a family, and we had to stand together, no matter what.

I had grown tired of seeing the destruction that the drugs were causing in our city and on my brother particularly. So, I worked with the police to try and end this game by providing as much assistance as I could. Upon sharing my concerns, the police officer showed interest in my proposal and we began strategizing. I knew that going undercover was the only way to catch this guy, and with the officer's help, we set out to bring the dealers poisoning my brother to justice. But, to my dismay, it seemed as if the government's policies were working in favor of the drug dealers. Every time they were arrested, they would end up back on the streets, free to continue causing chaos. I was appalled at the system's inefficiency and lack of accountability. The situation had become personal, and I was not going to back down until we made a considerable impact.

My brother was being poisoned by one drug dealer and this guy had even started making threats towards my father's home. It was at this point that I knew I had to take action. With a trembling hand, I dialed the phone number of the drug dealer responsible for my brother's decline. My heart racing, I carefully and humbly begged him to stop immediately. I didn't plead with him for myself, but for my brother who was exhausted and couldn't take it anymore. However, he replied, 'I will drug your brother until the day he dies. And if you try to get in my way, it will be payback for you and your family.' He knew critical information about me and my family, where I worked, etc.

The words echoed in my mind as I frantically dialed the police station, hoping they would be able to help me out of this nightmare. But when the officer on the other end told me that there was nothing they could do, I felt vulnerable and desperate. However, he cautioned me to keep my doors locked for protection purposes since this drug dealer was extremely dangerous. But I couldn't let them get away with it, and I knew I had to take matters into my own hands. With a determination burning within me, I set out to face the dealer, risking everything for the safety of my loved ones. It was a gamble, but one that I was willing to take, no matter the cost.

As I approached my brother, the gravity of the situation weighed heavily on my heart. The frustration and anger inside me churned, burning brighter with each passing moment, and I couldn't shake the feeling that we were all living in a type of warzone.

My heart raced as I thought about the man who knew too much about me. How did he know where I lived? What other information did he have? These questions swirled around in my head, taunting me with their unanswered mystery. I felt like I was being hunted, and that any wrong move could lead to disastrous consequences. That man, with his knowing gaze, had me cornered in a state of perpetual fear. I felt so helpless, with no escape from the sense of danger that hung over me.

I had grown tired of waiting for the police to take action and do their jobs. What if it was too late? What if the dealer already had his claws on someone else, I loved or cared about? It was a thought that plagued me, consumed me, and made every breath feel like a battle. So, I got behind the wheel of my car, drove without a destination, and let my emotions run wild. The waiting game was over. It was time for action.

When I finally spotted the dealer's house, I was overcome with ferocious anger and the need to act. I stormed into the house with my brother at my side, ready to confront the person who had been hurting our community for far too long. It was at that moment that everything changed - the feeling of power I had when I broke his existence is something that will stay with me forever.

I felt like I was in an action movie, going to once and for all take out the bad guy. Stepping in, I couldn't believe what I was seeing. The hollow, empty space sent chills down my spine, and the eerie silence was only made worse by the filthy mess that was left behind. Moldy boxes, overflowing trash, and dishes piled high in the sink only added to the unsettling feeling in my gut. The smell was horrendous. It was as if life had been sucked out of this place, leaving behind a haunting memory of what once was. But as I looked around, I realized the truth of the matter. This was a drug house. His girlfriend's angry shouts rang in my ears as she tried to push me back outside. My eyes scanned every corner of the room. The air was thick with tension; I could feel the danger lurking around every corner. But I wasn't going to back down. I was deter-mined to find him and put an end to his illegal operation. With every step I took, the stakes got higher, but I refused to give up until justice had been served. I ran upstairs to find the dealer in shock to see me standing there in his room. I lunged at him with a Superman punch and my fist collided with his cheekbone. I pounced over him and punched him until I could no longer breath. Blood dripped down his bruised and swollen face as he lay motionless on the ground, groaning in pain. His eyes were completely closed, and his nose was bent out of shape. As he lay there, I couldn't help but wonder if this was the end of his drug-dealing career. It was like a switch had been flipped inside of me, and all my rage and frustration came pouring out.

My brother, desperate to get his next fix, was scouring the drug dealer's house for something to get him high. I knew the risks of his actions, the potential for danger rooted deep in my bones. But my brother, fueled by his addiction, was blind to the danger. The situation was nothing short of chaotic. I had just finished giving a drug dealer a taste of his own medicine when his girlfriend rang the police. It was almost too perfect - the thrill of the experience had me laughing uncontrollably. I started to turn, ready to make my escape, when out of nowhere, the dealer's girlfriend appeared, waving a frying pan and swinging wildly in my direction. Her screams for the police only fueled my amusement. It was like something out of a movie - a comically tragic scene that I couldn't help but laugh at. But as I dodged the dangerously wielded pan, I knew I had to make my getaway before things got even more out of hand.

When the police arrived at my dad's door about an hour after the incident. I answered the door with my hands in front of me, in the formation of an arrest. The officers calmly asked me what I had done, my name, and if they could have a moment of my time. I obliged and informed them that I knew exactly why they were there. I had warned them that if they didn't help me, I would take matters into my own hands. And it seemed that's exactly what I had done. The tension in the air was palpable as the officers asked for more information. I couldn't help but feel a sense of adrenaline rush through me as I recounted the events that led up to this moment. The struggle, the anger, and the eventual explosion resulted in my rash actions. But now, facing the police, I knew that I had to take responsibility for what I had done.

One of the cops took me aside. He couldn't understand why I would admit to it since there was actually no proof. He said that I should have not told them the truth and they would have just laughed it off. I knew I had made a mistake, but I didn't want to lie. They said they would have let me go, but I had to be arrested since I admitted to my crime on his camera. It was a moment I would never forget, because the cops were themselves done with the mayhem this criminal had created and since he was beaten there was no proof, I did that. Nevertheless, since I had admitted the crime, I had to follow the protocol.

It was an experience I never thought I'd have to go through - spending a weekend in prison. The walls were barren and cold, and the air was dry and stifling. My small frame trembled with anxiety as I sat huddled in the cramped cell, alone with my thoughts. The iron bars were unforgiving, mocking my youthful innocence. I couldn't help but wonder how I had ended up in this place, a place meant for criminals and deviants. Tears burned at the corners of my eyes as I thought of my family. But at that moment, I knew I had to be strong.

I had to share a toilet with dozens of other inmates, and the smell was unbearable at times. The food we were given was bland and lacking in nutrition, leaving us feeling empty and unsatisfied. But perhaps the most difficult part was having to obey every single instruction given to us, no matter how trivial or degrading. There was no room for independence or individuality. Looking back, I can say with certainty that the experience of being in prison was not a good one.

When I got out, I was relieved to hear that I could dismiss the curfew since my job required me to work nights, but it didn't make the label of "criminal" any less uncomfortable. It was as if I was automatically assumed guilty until proven innocent. To add to the negativity, one of the officers seemed to take pleasure in bullying me when I would sign in. But looking back, the worst was over in just a few short days. I learned a lot about myself and my resilience during those few days, and I hope to never have to repeat the experience.

My sister, who resided on the west coast of Canada, 3000 miles away, came to visit for support during my family's dark times. Together, we searched for a place where my brother could receive the help he so desperately needed. It was heartbreaking to see the life of my brother slowly being taken away by what was perceived as a harmless drug by pharmaceutical companies.

It was a strange feeling to be both embarrassed and deeply concerned for my brother at the same time. Watching him spiral into the dark depths of drug addiction was devastating, but it was the public displays of his behavior that twisted my gut with shame. I wanted to help him, to shake him out of his stupor, but I knew deep down that change had to come from within.

I am a firm believer that change could only come from within. I knew that no external force could ever convince someone to make a change if they did not truly want it for themselves. It was a personal journey that required a deep level of reflection and commitment. I had witnessed many individuals attempt to change for the sake of pleasing someone else, but inevitably they always fell short. It was only when the desire for change came from within that true transformation could take place.

I'll never forget the relief I felt when my lawyer called to deliver the news that all the charges against me were dropped. It was like some kind of miracle. It had been a long and stressful ordeal, with the looming threat of a possible 9-year conviction weighing on my shoulders. But that phone call changed everything. Apparently, the drug dealer who had been responsible for all the trouble had been killed at a party, and under Canadian law, that meant that the case is thrown out. I was free. It was a surreal moment, almost too good to be true. But as I stood there, phone in hand, listening to my lawyer's voice, I knew that the nightmare was finally over. It was as if nothing even happened.

Chapter 3

Love Conquers All

After the door slammed shut behind me, leaving me to emerge out of my not-so-suitable shackles with a newfound knowledge of life, I was ready and motivated more than ever to take control of my future, which included finding that special someone. Little did I expect that after coming to terms with being done with bars for good that it would prove much more challenging to search for someone who could understand and relate to my journey in life than fellow inmates. Strewn with disappointment, even going for dates became harder than I thought — but I remained hopeful for the potential of finding someone who can appreciate something more meaningful... A genuine soulmate. Someone to whom I could relate.

Every meaningful relationship requires a level of relatability in order for it to stay alive and healthy. When we take the time to truly connect with others, it becomes far easier to empathize with their perspective, interests, and experiences - all of which is essential for sustaining relationships in the long run. I believe that this connection of relatability is what creates real moments of understanding between two people that can become strengths that are invaluable when faced with periods of instability or challenge. These shared connections turn acquaintances into friends and strangers into family through the development of deep empathy and appreciation for each other - a feat that cannot be replicated without deepening our sense of relatability in our relationships.

When I look back over my story of love, it's amazing to think that it all started with a great friendship - in particular, the friendship of a few people who thoughtfully decided to introduce me to the woman who would one day be by my side for eternity. To say we started off as strangers would be an understatement - neither of us could even imagine at the time what was going to happen between us. But life, and perhaps fate, had other ideas, and after twelve years together, our love has blossomed into something so much bigger than any of us ever expected. When I first looked into Samantha's beautiful blue eyes, I knew we had something special. Above all, I'm forever grateful to my friends whose suggestions gave me the beautiful gift of a lifetime. I knew I had found the person of my dreams.

As we prepared to welcome in the festive season, I had a very special surprise of my own an engagement proposal to my love. Knowing that one month later our daughter would be born, on January 9th, proved to be a momentous day not just for many around the world celebrating the holidays, but also for our little family as it marked a special new chapter in our lives. Reaching this milestone within such close timing really made it extra magical for us and we've cherished every moment since. It's been the most incredible journey and I'm glad I was able to start it off by proposing to my beloved partner.

I married my incredible wife, Samantha, with a secret blessing in tow - her eleven-month-old son Chase (my stepson). Our love was something magical and just months into our relationship, we were blessed again when I became the soon-to-be dad of a baby girl, we named Layla. God truly made me feel overjoyed by bringing us two beautiful children; it's an emotion that still feels surreal today.

Layla's birth marked a major milestone in my life and the start of an enthralling journey. As I flew around to work, unfortunately, it meant that two weeks at home was all I had to reconnect with Layla - who changed dramatically as she grew month after month. All thanks to God, a year later, I also had a son, Oliver. Oliver is our hard little worker.

Celebrating children's milestones is something every parent should experience, but alas, it felt like they passed me by quicker than ever before. That is when I made the decision to get ahold of a cell phone; this opened up umpteen possibilities for expanding my knowledge which captivated me beyond measure – leading forward on one very remarkable path indeed.

For the last two decades, my mining career has taken me around Canada as well as abroad. I had a unique opportunity to venture overseas and teach in Indonesia - which was extraordinary. It provided insight into different working processes while immersing myself in an unfamiliar culture, completely unlike anything back home here. Little did I know that this journey would be cut short when COVID hit just prior to my return.

Despite the illusion of equal treatment for all, I learned women experience a drastically different reality in many parts of the world. This injustice becomes even starker when set against places like Canada, where taxes are implemented to punish people who harm their environment, whereas, in countries such as Indonesia, this complicity from authorities is overlooked and often accepted by those living there due to strong cultural influence. It's time that we break away from our 'brainwashed' lifestyle here on Earth and start giving everyone—regardless of gender or location —the respect they deserve.

While away, my daughter asked me the heartbreaking question of why I had to work so far away from her when wrapping up my last work visit to Indonesia. At that time, due to my job, I was abroad for stretches of 10 weeks at a time and only home for six weeks. To explain myself, I answered with as much love and compassion as possible, something along the lines of needing extra money - little did she know it broke my heart too! Even though she was just seven years old then (and we have an incredibly close bond), hearing those words truly hit me like no other. But I had to stick to my bread and butter if I wanted to survive.

Believe it or not, in my storied mining career, I had a hand in unearthing and reburying bodies. Oh yeah, some crazy stuff happened... rescues galore! When you get down to the nitty gritty though what did I actually do? Well, picture this: A big underground corkscrew system that takes miners where they need to be - Yep, that would be me. Or one of those shafts with elevators thousands of feet underground– yep again; all expertly constructed by talented miners like yours truly.

It's hard to believe that, while I was working in the mining industry almost twenty years ago, companies would actually bid against each other based on the number of fatalities they predicted! It was as if human lives were worth no more than a few arbitrary points--almost like some twisted game. What made it even crazier was that these companies promised only five deaths and hoped for fewer. It's truly terrifying to think about how carelessly certain businesses used to approach safety two decades ago.

When it comes to building a mine shaft, safety is always the biggest priority. Mine work used to be much different – discussing how many fatalities were acceptable was part of the job. Thankfully nowadays, there's an abundance of laws and regulations in place that ensure worker safety. In fact, I've been doing this for twenty years now, and my success even allows me to take jobs around the world - did you know I once made 400k dollars working on a project overseas!? Talk about crazy money!

Starting from nothing, I powered through the challenges life threw at me and went on to become a superintendent of project consulting jobs despite my inexperience in this field. Although difficult times taught me valuable lessons - like when I got fired for speaking out too much - hard work across various projects around the world allowed me to experience even greater opportunities, such as visiting

Singapore, Hong Kong, and Korea. It wasn't easy but with sheer grit and determination gained along the way there was no stopping this determined journeyman.

My life was like a roller coaster ride, and I did what made me happy. My wife understood how dangerous my job can be and stayed at home and took care of the family, the house, and everything else while I pretended to be an all-conquering provider - making sure everyone had food in their bellies. Oh, yes! We lived well with a nice home, cars for everybody... because who doesn't love luxury goods? I once bought a 1949 Plymouth on a job in Vancouver Island. The furthest west of Canada. I drove it home to Ontario, thousands of miles across Canada with a good friend Chris Mets.

Besides, who says you need a fancy degree to be successful? Certainly not me! More often than not, I've seen those who put the most elbow grease into the job gets ahead of those with a head full of facts and figures. Grit – hard work, dedication, and resilience - is the secret sauce to maximum success no matter what your educational pedigree. So, roll up those sleeves and dig in; there's greatness to be made whether you have an MD or only a GED. And if nothing else, you'll show them you've got the real brains that really matter: the ones between your ears!

Chapter 4

China's Virus

It was the middle of December 2019 and I was traveling home through China. Little did I know that there was a very dangerous virus going around in the region, and it seemed to be spreading fast. As news started getting out and travel bans throughout the area were happening, I had a sinking feeling that I might not be able to get home as planned. This could change everything for me as a traveler, but also for people who already were infected by the virus or were about to be affected by it.

Though the coronavirus pandemic has had a devastating global impact, discovering the true origins of this novel virus still remains an ongoing mystery. While experts have theorized that it was transmitted from animals to humans (most likely bats) somewhere in China, which now we know was a lie. It never came from an animal. From what we know, one thing is clear - wherever it began, its reach has now extended far beyond its source and continues to wreak havoc worldwide.

To be honest, I had no idea about the virus until I got interested in politics. After that, I realized that some news organizations were being too tight-lipped about it. Despite some areas of the world, like

America and China beginning to report videos of what seemed to be a potential pandemic, it felt like media sources weren't doing enough to emphasize the danger this virus may present. When that information began coming out, I was quite shocked and concerned for those who had been unknowingly exposed to the virus for so long without protection against it.

At first, when I heard the news of the virus emerging in China in December, I realized I was there, I didn't take it too seriously, as sensationalized news often appears on screens. I couldn't have imagined it having such a devastating impact that would claim thousands of innocent lives across the world. Though conscious of some details, I assumed this too was one of those things that tend to pass with time, never expecting it to be anything out of the ordinary. Little did I know then what this virus was about to bring with it.

However, President Trump was making a strong case against Chinese carelessness and pointing out that their lack of caution posed a danger to other parts of the world. This made me sit up and take more notice since Trump always stands for American values no matter what. His words stuck with me since he was so adamant about holding China accountable for its actions. It just seemed so surreal at the time - a virus? - but here we are, grappling with its catastrophic impact upon our life's months later.

Politically-motivated comments about his every move seemed to dominate news cycles throughout Canada. The phrase "Trump is crazy and racist" would drop in conversations as if it were a fact, but despite all of this I knew something was going on behind the scenes. China had been reporting cases since December and yet mainstream outlets still denied the severity of it all - until finally, they could deny no more. I had to trust my gut that what Trump was saying was likely true, even if there seemed like external pressures preventing journalists from doing their jobs correctly.

The COVID-19 pandemic has changed many things in my life and one of them is how I view the news. Everywhere I looked, there were stories about fake news being reported as facts by what should have been reliable sources. Despite having fact-checkers present in a lot of the bigger news corporations, it seemed like almost every part of their reporting had some misinformation attached to it. It made me realize that there is often another version to a story, and blindly trusting any one outlet was no longer an option.

But it turns out actually later on that they were just actors. There was a whole team of people, who had been hired to make the situation look worse than it actually was, as part of political propaganda. It made me even more paranoid about the real threat of what was going on and only added to my fear and confusion about this virus. After doing extensive research, I must admit that my initial fear was somewhat justified. The virus was indeed a pandemic, and it was rapidly spreading around the globe. Moreover, scientists were still in the dark about much of its origin, transmission, and potential treatments. I felt like we were all being thrown into this huge experiment with no real understanding of what would happen next.

As time passed by, I found myself anxiously trying to stay updated on the latest news and developments. I was constantly reading articles, watching videos, and engaging in conversations with people around me about this very topic. It was my way of understanding the situation better and helping to inform others on how they can protect themselves.

During the pandemic, I found myself constantly sharing information online. I would bring up statistics from government websites in an effort to prove that the virus was not as serious as people were making it out to be and the absurdity of the SOPs (like 6-foot distancing). However, some individuals would question my credibility and ask if I was a doctor. Eventually, I started jokingly referring to myself as "Dr. Rylen" on Facebook. This inside joke caught on with my community and even people at the grocery store began addressing me as "Doctor." Despite not having a medical degree, it was amusing to see how easily people trusted me with information just because of a simple title change. The doctor was now my new pronoun.

What started as a curiosity has now become my passion.

As one who frequently flew in and out of Asia at the time, I can personally attest to how disconcerting flying through China was on the 28th of December 2019. So much uncertainty gripped the atmosphere as people voiced outrage about what had been reported in the "fake news" regarding President Trump. Despite this chaos, I knew that the media narrative of Trump being a liar, when it came to his claims, he wasn't a racist, was far from correct. As such, I felt an unwavering sense of peace as I passed through that chaotic airport—knowing that despite all the anger and frustration, truth and reason still prevailed over our great nation.

My experiences have taught me that sometimes the truth lies in places we least expect it. It's up to us to seek it out and be willing to go beyond our preconceived notions in order to uncover it. We must understand that often times what is shown on the news can be misinterpreted, so instead of relying solely on them for information, we should seek out multiple sources to really get to the truth. In the end, it is our responsibility to ensure that we are informed by accurate and reliable sources. It's no easy task but it can be done if we stay committed to learning the truth and spreading its message.

On my days off, I had a routine that I followed without fail. It involved waking up early, researching topics that I was interested in, drinking too much coffee, smoking too many cigarettes, and neglecting my health by not eating properly. But everything changed when I started making public visits. I was amazed at how brainwashed and obedient everyone was, unquestioningly following all the COVID SOPs without believing in them. The most mundane tasks became a hassle, as even going to a corner store to buy milk meant following the arrows on the floor. I thought it was all silly and pointless, and I rebelled against everything. I made sure that whenever I saw an arrow, I went in the opposite direction just to prove a point and compromise on my family time. Looking back, it was a crazy time, but it felt good to be a rebel.

The pandemic has also made me more aware of the value of communication and collaboration. In such uncertain times, it is essential that we are able to connect with others and share our experiences so that together we can find solutions and create a better future for all. It's only by working together that we will be able to overcome this pandemic and come out stronger than ever before.

In China, I had experienced a world of new cultures and customs, but I knew the time had come for me to return home. As soon as I got back, a job offer in Canada was presented to me. Without thinking twice, I said yes! It's such a relief to be working here - so close to home and family. I'm glad that I'm here, taking this next step on my career path in order to provide better security and financial stability for my loved ones.

However, it would be an injustice if I am not mentioning what that situation looked like. Let me tell you about my new job. It's truly a dream come true...if that dream involved dealing with my children's schools. Yippee! Not only do I have to deal with the usual school stuff, like homework and parent-teacher conferences, but now I get to battle it out with the school board over masks. Because apparently my kids are special and don't need to follow the same safety guidelines as everyone else. I mean, who needs fresh air and oxygen anyway? Just give us all a nice big dose of carbon monoxide and call it a day. As if that wasn't enough, I've also become a self-proclaimed expert on all things mask-related, scouring the internet for any study or article that supports my cause. Who needs a medical degree when you've got Google, am I right?

As a parent, nothing is more important to me than the health and safety of my children. So, when I saw blood dripping from their noses after they returned home from school one day, I went ballistic. Their masks were always full of mucus and very dirty. We would send them boxes of masks and tell the schools to make sure they are changed. They would sometimes come home with the same masks they left in the morning. My wife and I would explain that they needed to change the masks every hour, but they were too young. They didn't know better. It was a difficult decision, but at that moment, I knew that I had to act quickly to protect my children's wellbeing. And so, I made the tough choice to pull them out of school for the time being. It's challenging, but I know that I made the right decision to prioritize their health above all else.

Of course, that didn't mean it was without its challenges. My wife had to bear the brunt of the work, managing our children's education while also juggling her own responsibilities. But even through the toughest moments, she never once wavered. Her dedication and strength were inspiring, and I am so grateful for everything she did to make this past year as successful as it was.

With the outbreak of COVID-19 and subsequent restrictions, many people have been cooped up in the same place for months - however, that has allowed us to really appreciate the importance of being close to home. Being near family, who can provide us with emotional support, or our own space, where we can access all the creature comforts of home life, is essential during such a trying time. Particularly for those who had to relocate for work or university and were suddenly barred from returning home due to travel restrictions, finding solace in their local surroundings was beneficial when far away from their loved ones. Staying close to home also means having access to resources like timely medical attention without any delay, making it an even more extraordinary comfort during this pandemic.

When it comes to family, I've learned the importance of staying connected and appreciating those who are always there for us. No matter where we are in life, it is essential to remember that our loved ones will never leave us behind — they will be with us through thick and thin.

Chapter 5

My Beautiful Addiction

When I think back to the first day I heard about the virus. I remember the sense of panic that filled the air. Everyone had already started wearing masks, and anyone who ventured to the airport was not allowed to leave. Obviously, this virus was something the Chinese were feeling anxious to contain, as even my closest news sources were hesitant to discuss anything related to it. It felt like the beginning of something big - although, at the time, we had no idea just how far its reach would extend - and unknowingly, we had all stepped into a pandemic unlike any other.

After hearing about the origin of COVID-19, I wanted to research further into the matter. This was when I stumbled across a video featuring a female scientist who had worked in the Wuhan laboratory and eventually was forced to hide and flee the country out of fear for her life. During her interview, she explained that carelessness at the lab could have spread this virus. It was shocking to me that such an issue could occur within a scientific setting, especially one as prominent as Wuhan. To think that something like this had been going on right under our noses while we were unaware was almost inconceivable - but it was real. Hearing first-hand accounts from individuals involved made me consider the entire situation's depth and much larger implications.

The mainstream media kept pushing a narrative that Donald Trump was always lying about the virus and calling it racist slurs to describe the Chinese, which made sense only to those hell-bent on falsely accusing him. He called the virus "the China virus" at the beginning of January and then moved on to describe it as "Kung flu," both quite amusing statements in their own right. People were predictably outraged, as dramatic reactions seemed almost customary regarding news outlets controlled by the left. President Trump often stands alone in his beliefs, and he will certainly not back down when refuted by mainstream media sources who, let's face it, often follow an agenda of their own design.

On another occasion, when Nancy Pelosi threw her party in Chinatown to celebrate Chinese culture, it felt like the ultimate irony as President Trump had just closed all flights from China, attempting to protect his people from a potential health crisis. Pelosi's message was clear: "Trump is a racist." But she may not have considered that history will remember him for being one of the first leaders of any nation to take bold measures during a pandemic. So, it looks like Nancy Pelosi's party did little more than make the Democrats look petty and tone-deaf.

As someone who keeps up with the news and global events, I was intrigued by the method of recording when everything started to go downhill. How could we detect if the virus was taking over with no test? As my questions mounted, more research followed - and I suddenly found myself in a world of data analysis.

After two and a half months of this, my curiosity got the better of me and I decided to do some research. I started looking at the number of deaths in the month of February over the last 20 years - one of the lowest death rates on record. My initial reaction was a mixture of shock and confusion; how could this be? Pandemics were ravaging parts of Europe and even Trump had taken notice - where was this supposed to be showing up in our mortality figures? For some inexplicable reason, the number remained surprisingly low, leaving me wondering what could possibly be going on.

Investigating the death rate of Covid-19 during the pandemic was extremely puzzling to me, so I began taking screenshots of my findings while using Google. To my surprise, I found several government and public health websites that showed the age of people who had died from Covid to be above ninety. This is quite paradoxical considering Canada's average life expectancy is around eighty-two; therefore, justifiably shocking that so many people have been passing away due to Covid after the age of ninety years old. This lost me in complete bewilderment over how this could be true since people who cross ninety would naturally die.

In an effort to protect myself and the people in my network, I turned to our own public health websites for the facts. I wanted to make sure that nobody could argue with the sources, so I went straight to the Centers for Disease Control and Prevention, the FDA website, and other government resources. Having all this data gave me confidence when sharing information on social media - no matter what anyone said about its credibility - because I knew what I was claiming was factual.

Though I had done extensive research and collected many facts, this served to alienate me from my own family and friends. Rejecting the data, I had presented even though it was backed up by thorough investigation, they instead labeled me as a conspiracy theorist, someone not to be taken seriously. It was a heartbreaking conclusion that made it painfully clear how much some people were willing to ignore evidence in order to hold onto absurdity.

As much as I didn't believe COVID was a thing at first, it did make me more aware of what was going on and caused me to become more skeptical. Over the few three months, it became increasingly more concerning that the world seemed to be basing its understanding of this pandemic on symptoms accompanied by the dreaded daily news coverage in Canada of another death due to COVID.

The need to critically evaluate scientific evidence has never been higher than it is today. As the growing concern about infectious diseases and the increasing spread of misinformation have become prevalent, it has become increasingly difficult to distinguish between what is scientifically reliable and what is not.

Unfortunately, this has resulted in some people making decisions based off of symptoms rather than scientific evidence. Instead of calling for a professional to make an informed diagnosis, people are taking matters into their own hands and believing whatever information they find online without properly evaluating its reliability. This is not only irresponsible but also potentially dangerous as it could result in improper treatments or missed diagnoses. Science exists to provide reliable facts and should be used appropriately when it comes to understanding how our bodies work and reacting to medical issues.

It felt overwhelming that every symptom imaginable – sore eyes, sore back, coughing, runny nose, pinkeye, sore stomach, toothache– was being chalked up as a COVID symptom, causing Covid-deaths even if it hadn't actually been tested for. This pervasive virus quickly infiltrated our workplaces as we all had to answer questions every day, asking for every symptom known to man. On my final few days before widespread lockdowns were implemented, I still remember going to work early one morning and recollecting how different the atmosphere was from when it had all started just a few months earlier.

I remember going to work like a ticking time bomb, just waiting to go off. I had an itchy throat and I was really nervous walking through the gate of my work, thinking, 'Oh my god, how can I hide this?' The government was making all sorts of false promises, so though I had suspicions, I figured the fewer questions I answered, the better. So, whenever they asked if I had any symptoms, even if I did, I'd just lie and say no – there was no way in heck I was going to miss a day at work! As much as possible, I kept my mouth shut and just went about my business like nothing was wrong... as if anything ever is.

Shortly after march kicked in, safety measures were implemented, such as wearing masks and putting up plastic curtains every which way. I couldn't believe how restrictive it was! We were made to move like a herd of sheep - along paths and trails that had been prescribed for us. Even transportation was stripped back to the bare minimum - It all seemed a bit beyond what was necessary at first, but in hindsight, it was definitely necessary if we wanted to supposedly stay safe during these uncertain times.

It was an absolutely surreal experience to sit in a truck outfitted with plastic shower curtains erected between the benches. I mean, really? Were we supposed to trust these flimsy plastic walls as a barrier against a virus that had already wiped through countries like wildfire? It was almost comical how desperate we were to find a solution that would work. To top off the weirdness, this so-called COVID team was rapidly indoctrinating us with their claims that the curtains would protect us from any harm. Seeking an answer amidst all the uncertainty, it seemed like we were grasping at straws - hoping that if we shook them hard enough, something might come out of it.

I was absolutely outraged when I heard about the masking and six-foot distancing rule. I mean, come on! To put something like that in place for a virus that could travel for miles? It seemed ridiculous. Then I looked into aerosol viruses and realized these tiny particles were 30 times smaller than a single hair, so of course, a mask would do nothing! After looking into it further, I found this study by the Brownstone Institute, which found more than 150 comparative studies and articles on masks and their effectiveness. I was flabbergasted: how could all these prestigious universities and laboratories be getting it wrong? That too in 2020?! What the heck?!

People usually go about saying: "Really? How could a face mask be bad for me or anyone else?" It's understandable why some may think wearing face masks would have an impact on controlling the spread of COVID-19, but unfortunately, the research shows that there is little to no evidence that surgical masks are effective. In fact, there are even cases in which they can be detrimental as they cause hundreds of injuries. With all of this in mind, it becomes clear that wearing a mask is more of a safety ritual than a necessary precaution to stop the virus's spread. Needless to say, it has been confusing now having to get used to carrying around a mask everywhere when going anywhere!

Besides, my city was a refugee of fear: all you could see everywhere were double masks and worried faces. We had been strong and united before, but now it was as if we had all been caught up in the powerful tool of divide and conquer. It's an ancient technique of rulers from centuries ago, meant to weaken us by turning us against each other. Even my own family couldn't resist this – I faced scorn for merely questioning what we'd suddenly found ourselves living through! Honestly, who knew the pandemic would have such a huge dark side and to such a great extent?

Going shopping with me must have been quite something for my poor wife! I think she was mortified that her husband kept insisting on wearing paper bags, Halloween masks, and whatnot to make a public statement of our silliness. I'm sure people at the store thought I had gone mad; yet, I knew it wasn't foolish at all, but rather an important show of resistance against the status quo of conformism. My wife had to learn that sometimes you simply need to stand up and be silly in order to make yourself heard - although preferably without making your spouse look ridiculous in the process!

Rebellion can come in many forms and for me, it was wearing a paper bag or plate over my head with two eyeholes cut out. It's funny how I could walk around malls and people would try their best to get me kicked out, but security wouldn't because they saw that my face was covered- so far as they knew -I was compliant. I even got fancy at times and drew funny pictures on the paper plates and other masks I would wear; it was definitely an interesting fashion statement! The fact that I showed up to places wearing masks before the pandemic just goes to show that you don't need a crisis to showcase rebellion.

I was extremely bothered when I discovered that the medical bureaucrats started advertising a remedy vaccine soon to hit the shelves. It was pushed on mainstream media at first as, one shot and done. It shocked me when I learned about this clinical trial for the COVID-19 vaccine. How had they managed to pull off something that had never been done before?

It seemed crazy, and yet Pfizer's own documentation showed that although the effectiveness of the vaccine was not 100%, world leaders still hyped it up like it was a heavenly antidote. Even still, stories of people who got vaccinated coming down with COVID felt like proof that it wasn't all that it was made out to be; a bit ironic if you ask me. What's more, everyone around me was so insistent on getting this vaccine as soon as possible without asking questions or considering what could go wrong. While I think having one available might be a good thing ultimately, I'm wondering how did we even get here.

It almost seemed as if these large pharmaceutical companies were claiming to have a magical 'cure' that would save the world, but it turned out that these vaccines actually did almost nothing at all! What boggles my mind is how they were ever able to convince medical professionals and everyday citizens that these random clinical trials would miraculously resolve our global issues with just one injection.

I felt like I couldn't win - first, I was labeled an "anti-vaxxer," a conspiracy theorist, and an idiot by my coworkers for no apparent reason. Time after time I was put in corners, and made to feel as if I had some kind of infection that no one else wanted to be around. Instead of doing anything about it, my manager would just make sly comments such as "he hasn't had his vaccine yet," implying that there was something wrong with me for not having done so. Eventually, the man responsible for these comments ended up getting removed - but not before making me endure months of being treated deplorably. After being exposed to this kind of behavior on a daily basis, it's now amazing to think back and recognize how much worse things could have been if a change hadn't occurred.

Despite the lack of scientific evidence, authorities still implemented masking and social distancing during the pandemic. At the time, they were operating with no evidence and no concrete conclusion...pretty much thought, 'Why not give it a go?' It's funny how sometimes our most daring decisions end up without any solid foundation or backing.

To strengthen my point, let's take Project Veritas into consideration. It shook the headlines with its latest exposé - an undercover interview revealing that COVID-19 is nothing more than a cash cow for Pfizer. In what seems like a conspiracy right out of a Hollywood movie, the director of Pfizer actually confirmed these accusations! He went as far as to say that they're making viruses and vaccines and will continue to do so in the future. To make matters even worse, platforms such as YouTube try to cover up these incriminating stories by taking down the videos. Seems like "freedom of speech" isn't something they take seriously...

The global reset agenda they have planned during this pandemic is totally absurd. Make no mistake. It's not so that we can really pull ourselves out of the crisis but more just so that some people can make tons of money! At least our Prime Minister was honest when he said this was an ideal time to implement it. So, I took it upon myself to investigate what they had in the works, which is almost comical. They seem to want us all to be penniless, give up farming and eat bugs, and above all, keep tabs on every single thing we do. What a world!

Despite what these so-called authorities think, we're not oblivious to their attempts to control us down to our movements. We saw it loud and clear when they tried to shut down the truckers' rally and froze out all donations from opponents. They believe themselves to be great rulers, pulling strings and issuing decrees as if they are mere mortals above us. Emergency orders? Please. It is quite clear that it's just an excuse to prop up their organizations while bankrupting our countries in the process - this isn't about public safety but rather, power and control. We haven't been living in a fairytale; we are all now the peasants in their kingdom who must bow down to their whims or else face further injustice. How lovely.

With all of the Covid lockdowns, I figured it was time to get creative when it came to having a good time. When the government said not to leave our homes, I saw that as an opportunity for adventure! I invited everyone over and we had a big party - after all, we weren't going anywhere else, so might as well have fun! We also made sure to take advantage of every loophole in the rules - when they said we couldn't leave our homes at Easter time in 2021, I took that as my cue to buy a camper, truck, and boat all in one week so I could still go on camping trips. Talk about making lemonade out of lemons! Defying the restrictions actually ended up being one of the most memorable experiences ever; in fact, I'm pretty sure this was the best year of my life, thanks to Covid regulations.

During Christmas break, the government told us we couldn't get together for the holidays. I called the city to ask how many people I could rent my house out to. The city didn't understand the question and said she didn't know. She said that there is no number. I told the girl that I will be renting my house out for New Year's so I could have a birthday party for my friend Skylar. She laughed and told me I was a genius. So, I took it upon myself to advertise my house for rent by placing posters on the windows with available signs. I made rental agreements for my guests to feel welcome and accounted for, whatever the law was saying. I did some research and learned that as long as they were all living here, I was perfectly within my rights! I guess you could say – in true sarcastic fashion – that I was ready for the cops at any moment because hey, that's what you do when you're a responsible landowner, right?

At the end of the day, it's important to remember that we are all in this together. We can rise up against oppressive systems and take back our individual rights without sacrificing public safety. It starts with us – standing up for what's right and not letting fear rule over our lives. So, let's continue living life on our own terms, shall we? Let's keep standing up for what is right and stay informed on the issues that matter most. Together, we can make a difference in our communities by speaking out against injustice, no matter how big or small. We have the power to create change – now let's use it!

Chapter 6

The Fringe Minority

After about nine months in Canada, the mine I worked at went on strike. Me being a contractor meant that now I was looking for a new job. It didn't take long to find myself ready back in at another mining project close to home. I was back on the dreaded probationary period! At least by the end of that three-month "sentence," I could begin to assert myself with confidence, despite my newfound appreciation for certain existing practices. Ah, what it means to be a grown-up!

Working at my new job was rough– The morning questionnaire created rage right off the bat. I mean, don't get me wrong, safety is super important, but now knowing that there might be an agenda, made me not want to answer a dam question.

People with symptoms ranging from a headache to a cough suddenly found themselves second-guessing their health. This was because it felt like virtually anything could potentially be classified as a symptom of COVID-19. It became an issue because saying that you had a minor health issue could lead to you being flagged for a potential COVID-19 case and the ensuing quarantine and testing procedures that followed. This created a great deal of anxiety amidst an already chaotic time. So yeah, that really killed my motivation in the mornings.

Above that, Oh joy, what luck! Canada was very late providing PCR testing compared to a lot of countries. Way later than Trump's America. And then, low and behold, when we finally did decide to take the plunge and use the test, we discovered that the package has a handy warning label right on it: "does not test for COVID-19." Well, isn't that just fantastic? It's almost like we wasted all that time and energy, not to mention money, for absolutely nothing. Thanks for nothing, the testing industry.

Every morning, I was greeted with the ominous thermometer gun pointed at my forehead. It seemed like overkill to me, and after a while, it even started to feel a little bit scary. I finally told them that if they wanted to take my temperature, could they please point it at my wrist? From then on, things got even weirder, as there were signs and posters everywhere about COVID seemingly plastered on every surface, every couple of feet. How come our teachers never make us learn about pandemics in health class? Then again - can you really teach common sense?

My fury regarding Covid was on the brink of uncontrollability. Everywhere I would go were signs warning of safety precautions, making me downright sick and inciting me to rip them down out of sight when no one was watching. Even at my new job, everyone was brainwashed by the government's stance on the matter. I felt so disconnected; until I met James. He made quite an impression on me in our first meeting, and we gradually became friends - it felt nice to have a sense of normalcy again!

When James and I met, it was like a breath of fresh air for both of us. Here was a man who was so conservative and religious, so steeped in tradition that he kept so much of himself hidden from the world - but he found himself instantly intrigued by someone who wasn't afraid to ask questions and use this strange time of COVID as an opportunity for learning. From that point on, we got close very quickly. It didn't take long for me to realize that despite our differences, James had a brilliant mind for understanding people and connecting with them on a personal level. We began spending time together, and our relationship blossomed into something truly wonderful.

When I began my job, James gave me lots of advice about presenting myself, like staying calm and showing love and understanding. He was so passionate about it; I felt like he was a long-lost grandfather who wanted to impart wisdom! Well, I followed his advice, and the results were astounding.

At the time, it was January 2021, on the internet I was seeing this group of truckers, calling on all truckers to join in what started a movement that rocked the world. Weeks ahead of the massive rally, rumors of their route to Ottawa spread like wildfire. After dealing with these mandates and what they meant for their livelihoods, there wasn't any doubt that they'd take a stand– or an entire drive– in solidarity against what they stood to lose.

The story of the truckers' revolt against government vaccine laws began with just one man's plea for his basic liberties - only to lead to an immense wave of outrage from truckers all over the country. What started as a petition quickly turned into a full-blown revolution, with frustrated truckers objecting vocally to what they saw as a grotesque violation of their rights and freedoms. It was truly a seismic event because while they had already been struggling against such regulations, this seemed to be the final nail in the coffin that pushed them over the edge. From then on, it was clear that there would be no stopping these passionate individuals in their quest for justice.

They said it was going to be big, and they weren't wrong. News of the rally traveled like a raging blaze down the Information Superhighway, filling inboxes and popping up on newsfeeds until everyone had heard it. From British Columbia, all the way to Ottawa, three thousand miles away, and over the course of weeks, enthusiasm for the mission grew with flags and greetings appearing in every town along the way. It could have been a scene straight out of a movie if anyone had scripted that far ahead! In any case, citizens from across Canada stood hand-in-hand in an effort to hold their government accountable - now that's something worth shouting about. Surprisingly enough, the fake news suppressed this story completely. The lame stream media said the truckers were rallying against road safety. What a joke.

Before long, emails were flying and headlines screamed about what was about to happen. My home is near the highway, and I thought of a brilliant plan, asking the owner of a nearby motel if we could use their parking lot for our plan to greet the convoy.

On the appointed day, a miraculous event occurred. Hundreds of trucks loaded with supplies set out on a remarkable mission, heading down highways and country roads in an organized symphony of activity. It was impossible to miss the incredible energy radiating from the convoy, a sense that they were leading something monumental and momentous. Everywhere they passed by, people gazed out in awe and admiration at the powerful sight of all those vehicles setting off on their task. It was truly impressive, and there's little doubt that this will be remembered as one of those days never to be forgotten.

It was a cold and chilly day in Canada, the coldest month of the year, to be exact, but that didn't stop our family and friends from coming out in full force. Everyone showed up to the motel with their dedicated flags and braved the cold despite the temperature. I remember feeling grateful for having hot pads to put into gloves for the kids; it was like nothing else could have saved us from that numbingly biting chill. Luckily, we had enough hot cocoa to go around, so everyone was able to drink some warm and tasty liquid as our event drew to a close later on. In short, it was a fantastic event, and everybody was a true soldier on this frosty winter day.

A chill filled the air as we waited for the long-haul truckers to pass by. I could feel anticipation and unease massing slowly among us, but when the first hulking vehicle came into view, something in my body shifted. Hope swam easily through my veins, calming my heart and mind. My family members grasped together hands and gazed on with excitement and raw emotion. Even my usually stoic children brought tears to their eyes: sparking a warmth and comfort, I hadn't felt in weeks. We had no certain knowledge of what was to come next in life's journey, but one thing was undeniable - the truckers gave us hope during an incredibly difficult time.

For many of us, it was an unprecedented feeling of admiration to witness the event unfolding in front of our very own eyes. Seeing a small figure slowly approach alongside an enormous truck, with headlights shining like beacons of hope, felt almost supernatural - like a scene taken straight from a fairytale come to life. But a fairytale prince this was not; these extraordinary individuals were one of us; truckers who came to our aid when we needed it most. With their sheer determination and commitment, they were able to make all the difference: saving us from certain despair and providing solace amidst the darkness.

My family and I witnessed something truly remarkable, an astonishingly long convoy of trucks. It started when the first truck passed by our house and continued for hours. The convoy nowhere near ended before my eyes were burning from staring. Reports began to spread afterward that this convoy was thirty miles long, breaking multiple records for the longest group of people traveling together. Before this incident, the longest parade or convoy ever recorded stood at just three miles long; thirty miles outshone it easily, shattering all preconceived notions in its wake. It was a breathtaking event that will surely remain etched in my memory for years to come.

The government had pushed people to their limits, and many responded by standing together in protest. Most of the protesters were already vaccinated, but they could no longer stand the control over their medical choices. Emotions ran high as many of the protestors grappled with their newfound conviction. I will never forget a moment when my daughter followed me to the convoy. She was playing around with other children and remained curious when she noticed I was getting emotional. She peaked her head out from behind a vehicle just to get a glimpse into my feelings, likely not having ever seen me cry before. It was such an overwhelming experience that it is still hard for me to talk about it today.

I took advantage of this enchanted moment, taking the day off work to appreciate the convoy in all its glory, and offered James to do the same. We both felt this powerful urge of hope, something that was unexplainable - a rush of emotion so strong that it will stay with us forever. Despite having one last shift left between us, we paused that moment to take it all in. It was beautiful and inspiring, setting us up for an unforgettable day.

The next day after work, I excitedly informed my family that I was going to Ottawa with James. Knowing the media had already begun to distort our message, I wanted to do whatever I could to protect this movement and ensure its success. Despite having just finished a long shift at work, I felt invigorated and couldn't wait to leave for our journey the next morning. Hopefully, with two voices instead of one, we could make a bigger difference and help bring about some much-needed change.

When we arrived in Ottawa, our militant friends had already been there from the start. They relayed their heroic doings from the day so far, they told us about how they spent their time; spreading sand on icy roads and sidewalks tidying up areas, and helping prepare meals for later in the day. But more important than any of this was that everyone was unified in their mission, to protect the cause and not let it be tainted by pressure from either corrupted governing or media outlets determined to bury the truth. It didn't take long for us all to join forces and stand against injustices together.

As I arrived at the location, my mother was overcome with alarm and concern over what she had seen on the news. She kept calling me in an increasingly panicked voice, distress evident in every word. In order to console her and prove the false nature of this media portrayal, I had to explain in detail how stories can be exaggerated or even substantially altered. It took some effort from my end to help her see that all this 'news' was a lie; people had not been arrested, put in jail, or fined - if it weren't for me taking time to set the record straight, she would have believed it utterly. It felt as if I had foiled a massive conspiracy theory, liberating her from its treacherous grasp.

I made the decision to video call my mother, despite the chaotic atmosphere in the area, and show her the groups of police officers I had surrounded myself with and dared to walk up to them and hug and shake their hands. You could only imagine my mother's reaction when she saw this compared to what watched on the news that evening - shocked beyond belief at what she thought was real-life chaos. But I quickly tried to ease her worry by assuring her the news was manipulating footage and lying, simply because they are a 70 percent government-funded organization.

It was like witnessing a huge and incredible force of nature, as millions of people from all over Canada came together to participate in the event. As they worked to keep it clean, cook and set up activities, we managed to transform the space into a carnival-like street for the children. Here, snow hills were created around them so that they could play with joy and enthusiasm, different genres of music were playing at every intersection and people were encouraged to dance and be free. What the news was portraying to be chaotic, suddenly morphed into something beautiful; it was both surreal and magnificent.

During my brief stay in Ottawa, I was witness to some beautiful things as the indigenous people there set up cultural areas and hosted hockey tournaments that featured a special Stanley Cup. The vibrant energy of this special time prompted me to begin live streaming on Facebook so that more people could have the opportunity to connect with what was taking place. To my surprise, the audiences grew exponentially as they found solace in getting direct accounts that differed from what they were hearing in the news. For a while, it felt like being back home among family - gathering together and everyone fiercely devoted to carving out joy and celebration.

The first weekend in Ottawa was an eventful one. My friend James and I had traveled together and were making videos, calling out to miners, friends, and family to join us in contributing to this beautiful movement. I remember feeling sad when James had to leave because of something that happened in his family. However, the day that I was leaving, the moment was quite awesome. As I was preparing to say goodbye, two familiar faces tapped me on my back. It was my good friend and my cousin. They simply said, "We're here, man. You told us to come, and we're here." I was moved by their response and heartened by their support. It was a great reminder that when we call out for help, sometimes the people we want most show up to be the support we need.

Lost in a sea of people, I never thought I would stand out. But little did I know, fate had other plans. Amongst the millions, I was found. And not just found, but welcomed with open arms. Emotions ran high as I asked how long they were staying. Two days was the answer, and without hesitation, I asked if I could stay with them. My friend had to leave, so why not? The answer was a resounding yes. And just like that, my life took an unexpected turn. I was hooked on journalism via my live streams, and the energy of it all consumed me. The opportunity to spread the truth was just too good to turn down. I may have been lost in that crowd, but I had found my calling.

With millions of people around me, not wearing masks, it felt like a breath of fresh air. To celebrate this newfound freedom, we even took the bold step of visiting cafes and restaurants that were open and willing to take a stand. It was an incredible experience that I'll treasure for a long time. It was great to see these small businesses open making so much money their tills were overfilling. The mayor threatened anyone that opened their businesses with massive fines, but the people of Ottawa were also fed up with their agenda.

As I made my way home, something inside me ignited. It was a voice that said I had to stand up for myself and for my colleagues. And so, I made an announcement to my fans and to anyone who was watching me. I was going to do something at work. I was going to make a difference. I called James that night, the day before work, and I told him about my plan. I told him that I had written a speech and that I was going to do a work refusal. It was something that I had been asking for, for the last eight months. My concerns were never answered, but I knew that I had to speak up.

I can still remember the fear in James' voice as I told him what I was going to do. But his non-confrontational nature couldn't sway me. I knew the risks, I knew I could lose my job, but I had to stand up for what I believed in. As I told him my speech, I could feel his emotions threw the phone. It was as if he knew that I was going to win this battle. Even though he warned me that I could lose my job, I couldn't back down. I had to follow through with my plan and fight for what I thought was right. And when I did, the feeling of victory was indescribable. James may have been scared for me, but his support was what kept me going.

Chapter 7

The Lonely Fight

As I walked through the door of my workplace the next day, my heart was pounding in my chest. But I knew that my voice was important and that I had to be heard.

As the meeting time drew near, I could feel my nerves beginning to fray. I had brought up my concerns time and time again, but it felt like my questions were always being swept under the rug. This time, I was determined to be heard. I wanted to audit the system and use it against COVID, and masking.

The adrenaline was pumping as I got ready telling the guys in the change room to record me in the meeting. They knew I was up to something, and they all just chuckled. I loved my crew, and for the first time, I had the perfect crew. I felt a rush of excitement knowing that something big was about to happen. Being the temporary leader of the group at the time, I was responsible for hosting the morning meeting.

When the meeting began, I stood there guns a-blazing knowing calmness is my only option, my men realized what was happening, but the others seemed taken aback. I began the meeting by reading my speech asking for no interruptions. I declared that I was doing a work refusal, a necessary step when questions are left unanswered according to the green book. It was then that a manager interrupted me, grabbing my wrist and holding it up for all to see. "Look," he sneered, "Rylen's wearing a wedding ring. How can he possibly care about safety?" My blood boiled at this insinuation even though I had asked everyone not to interrupt me.

Imagine being in the midst of a passionate and invigorating speech, only to have someone try to interrupt you grabbing your wrist. Something a parent would do to a misbehaving toddler. That's exactly what happened to me. I had explicitly told everyone not to interrupt me, yet this manager, Tim, thought it was okay to try and stop me. To make matters worse, I had to wear a mask. So, when Tim tried to hold my arm up mid-speech, I ripped my hand away and scolded him. I told Tim that I had listened to his "bullshit" for months, and now it was his turn to listen to me. It was an emotional moment - one that I will never forget.

It was time for my voice to be heard.

As the words flowed from my mouth, the energy in the room intensified, and I could feel the eyes of my crew on me. This was my time to shine, and I was determined to make it a memorable one. The room was silent as I continued my speech, revealing the culmination of my hard work: paperwork, a notice of liability, and an ultimatum. Either provide the study or I am no longer wearing the mask. With conviction, I shed my face diaper and tossed it aside, taking a bold stand for my beliefs. Despite the fear radiating from my coworkers, I invited them to stand with me and breathe freely. Their eyes darted to the floor, avoiding my gaze, surely thinking I had sealed my own fate. But no, I stood my ground and spoke my truth. It was a moment of power and liberation, and I knew that no matter the outcome, I had made the right choice.

The air in the room grew thick as silence fell upon us. We were all waiting for someone to speak up, to break the tension. Finally, one of my managers did. But it wasn't the response we were all expecting. Instead of addressing the issue at hand, this manager took a cheap shot at me. He asked the group if anyone even agreed with me. He told the group he liked masks. I told him I was happy he liked masks. I told him I hoped he wore one when he went swimming. To show he was with me, James spoke up. He didn't get into the argument, but he made a powerful statement nonetheless. I'll never forget his words: "Someone needs to get Rylen a wheelbarrow because he's got balls the size of Texas." It was both hilarious and empowering, and it stirred up emotions that had been lying dormant.

As the manager hurried out of the room, I couldn't help but feel a sense of nervous anticipation. What was going on? James wanted me to put the mask back on. My thoughts were quickly interrupted by his urgent whisper, "Rylen, put the mask back on. This is what they have against you." My heart skipped a beat as he explained the consequences of defying the rules. I didn't want to be kicked out, but I also didn't want to be forced to wear that suffocating mask again. As I reluctantly covered my eyes, I braced myself for the worst. But nothing could have prepared me for what happened next. The manager stormed in with a team of security guards, eyes scanning the room until he spotted me. "There he is; he's not wearing a mask!" His accusation hung in the air, heavy with tension. And in that moment, I knew that everything was about to change.

When I turned around, I was wearing a mask. It was so sudden, so unexpected, that I didn't know how to react. I looked around and saw Tim and the security guard who was supposed to be keeping an eye on things - standing there, looking at me like he had no idea what was going on. 'What's happening?' Tim told me to go pack my gear and get off the property. He shouted at the security guard to remove me at once. That's when everything started to unravel. The guards told the manager that they couldn't kick me out because I was wearing my mask. Tim started freaking out, insisting that I wasn't wearing it before. And then things got really heated. I tried to tell them that Tim had put his hands on me and that I wanted him removed, but no one seemed to care. And then, right in front of everyone, he pushed me out of the room and slammed the door in my face. It was a terrifying experience and one that I'll never forget.

My heart was racing as I stood outside the closed door. I could hear Tim's voice, muffled by the thick wooden barrier separating us. During my speech, Tim said, "Hurry up, we don't have time for this." So now it was my turn to give this idiot a taste of his own medicine. I opened the door, my head poking into the room like a curious rabbit emerging from its burrow. Inside, I smiled at the group of men, all still white as ghosts, shocked at what had taken place. But my focus was on Tim. "Let's go, Tim. We do not have time for this." I said firmly, trying to mask the frustration in my voice. He had been interrupting me all through the meeting, throwing me off my game. It was time to wrap things up and move on. Tim slammed the door shut in my face, causing me to stumble backward. But I wasn't deterred. I knew I had to stay strong and focused, especially when dealing with someone as difficult as Tim.

I started reciting the Green Book when suddenly Tim charged out of the room. Panic seared me as I followed the next step - finding a safe room and getting a representative. I begged the safety director for help, but he seemed unsure of what to do. I coached everyone threw the process, and eventually, two managers took me into a room, but with no representative, completely disregarding what the book had advised. James, my colleague, could see my distress and told me to document every moment moving forward. I begged him to stay with me, but the management told him to go to work, despite it being against the law. It was a moment of sheer desperation and isolation, and I felt completely alone.

I can imagine the scene vividly in my mind as if it were yesterday. When all the men went to work, I saw this concept I liked and I couldn't help but give a high five to all the guys. To my pleasant surprise, they reciprocated with enthusiasm, but that was not the end of it. As they marched out triumphantly, heads held high with our silent conspiratorial gazes locked, the managers could not hide their fuming fury. They dragged me into a room where they threatened and attempted to bully me into submission, but I stood my ground. Armed with a pad of paper, a pen, and James' advice, I confidently informed them that the opposite of what they predicted would happen. The look of shock on their faces as they realized they stood to lose much more than they had bargained for was priceless.

Picture this: You're standing up for yourself in a room of higher authorities who just don't seem to get it. You want what's right, you want what's fair, and you want to start this process. You tell them that the Ministry of Labor needs to come to the site and investigate the work refusal. But instead of taking you seriously, they just laugh and brush you off. You're not one to be discouraged easily though. You sit down, notebook in hand, and start documenting everything. Every word they say, every action they take. And when the manager tells you to stop writing, you dare to write that down too. You refuse to let them silence you. They may think they have the upper hand, but you'll show them that the more they talk, the more evidence you gather.

After an hour of coaching these bullies, they looked at me and said, "Rylen, what do you want?" It was then that I knew things were going my way. It was then that the process started, and at lunchtime, the ministry officers would be there. They started checking on me every half hour, bringing me treats and gifts as if trying to make amends for what they had done. It was clear that they were feeling guilty and knew they were in trouble. I couldn't help but feel a sense of satisfaction knowing that I was the one who had brought them to this realization.

It was a daunting sight to see the ministry officers march in. For an entire year, they had been on a mission to impose fines on anyone who dared to disobey the mask mandate. So, it was no surprise that I felt a twinge of unease when I watched them enter the room. I knew that they weren't my allies, but their involvement was part of the process. And so, it began. My heart raced as I locked eyes with the ministry officer strutting towards me, his big build making him all the more intimidating. Even his mask was taped tightly to his face, as if he were getting ready for battle. I admit, I didn't have high hopes for the encounter. I couldn't help but think to myself, "Well, this is going to be a real challenge."

So anyways, he asked me what my concerns were about the masks we were all wearing, and when I told him, he had no response. It felt like he was minimizing my worries, so I kept asking him for answers. After all, I had the right to know what exactly we were breathing in. But instead of providing me with information, he required me to prove the legitimacy of my concerns with a study. It felt like he was putting me on trial, and I couldn't help but feel frustrated and helpless. Even worse, the Ministry then informed me I had 30 days to prove wearing masks all day was a safety concern. It was a heavy burden to bear, and I couldn't help but wonder why those in power weren't more concerned with our safety.

As I left the meeting with the Ministry, my mind was racing with thoughts of what was to come. Would they take my concerns seriously or dismiss them as trivial? The next day, I returned to do the re-Covid training, but this time with a newfound sense of hope. The Ministry had issued orders to the mine that they could not harass or intimidate me for the next two years. It was like a weight had been lifted off my shoulders. I was a worker who had stood up for safety, and now I was protected by the Green Book.

Doing the re-COVID training was a terrible experience, but without it, I would have never seen the company policy that required six feet of distancing and the company-mandated masks, even when no other people were not present. Even though the company had a policy, they were changing it when it suited them. This, I felt, was an entirely unnecessary measure that focused on tyranny rather than safety, and I knew that if I ever got in trouble for not following proper protocol, at least I had this precedent to point to in my defense. Keeping this "gotcha" moment at the forefront of my mind allowed me to be prepared for any situation and to ensure that I was following the guidelines that actually mattered instead of wasting effort on something nonsensical.

When faced with the question of what to do next, I told my boss that the day had been very hard for me. I was stressed out, having gone through physical and verbal assault, and needed some time to level out. I suggested that it would be best to stay home and start organizing what needed to be done next. Much to my surprise, one of the higher-ups came forward and offered his support discretely. He reminded me not to let others know that he was helping as he wanted to remain impartial. Although there were plenty of other people reaching out offering help, I chose to rely on this friendly face in disguise who could provide some guidance from within the organization.

I knew that if I didn't take action, no one else would, and my boss's behavior wouldn't be properly addressed. After asking him what the consequences were for putting his hands on me, I decided to take swift action by calling HR. Luckily, they believed me and partnered with me to initiate an investigation. They showed their support and genuine concern by immediately bringing in two other managers to interrogate my account of the incident. It was extremely satisfying to know that my voice was being heard and that appropriate action was being taken.

After conducting a thorough investigation into the workplace dynamics, I was approached by my department with a solution. They offered to relocate the individuals involved if I was no longer comfortable working with them after what I had been through. Knowing that this would allow for a positive work environment for everyone, I accepted their proposal, and the persons in question were subsequently sent packing. In doing so, my department showed me care and concern, leaving me feeling supported during a difficult time.

After the head engineer got removed from his position at the mine site, everyone assumed there wasn't anyone left to lead. People joked that, for once, no one could tell them what to do or make them complete mundane tasks. It was an oddly freeing feeling to have no bosses around to make sure people followed the rules and regulations. Although it was only a few days before a new set of bosses came in, it was fun while it lasted. Everyone had a good laugh and enjoyed the short break from having someone looking over their shoulder all day.

The challenge had been presented before me, but I was determined to stand my ground and protect the movement. After much consideration, two days following the work refusal I had committed to, I returned to my usual work week with a plan calculated in mind. Arms linked with my wife Sam, we decided to leave our hometown and take a trip together with our two kids and Sam's aunt Nannette and uncle Ricky.

Thankfully, our search for accommodation wasn't too hard; we found a great hotel close to the action and plenty of fun activities. With some strategic planning, we made sure to stay limber by taking regular warm-up breaks, while also managing to take full advantage of all Ottawa had to offer.

Our destination was full of surprises - there was a pool at the hotel, which made the kids beyond excited. Bringing my wife's uncle and aunt along with us was great too. They were an older couple, but nevertheless, they had an exciting energy around them. The second weekend of our stay meant more fun because even more stages with a lot more music were added to the already breathtaking scenery. We could feel so much love going on at that moment! There was definitely an atmosphere of joy that kept us engaged throughout the entire trip.

There was so much free stuff and activities everywhere around free toys, souvenirs, t-shirts from local companies, gloves, BBQs, and food. Things started getting even more exciting when a group of bricklayers decided to make an outdoor pizza oven in the middle of the street where people could cook whatever treats they wanted. The atmosphere was amazing, with everyone gathering by the oven and sharing stories while enjoying great company and the delicious smell of baking goodies. We truly had an incredible weekend.

During my time in Ottawa, I worked hard to stay organized and make the most of my time. Aside from exploring the city with my family, I was also taking steps toward furthering my research and setting up visits. While doing this, I knew that I had to be responsible and figure out a way to get my work to finance my doing this. So, I contacted my Head of Safety at the mine site and requested help. My company agreed to allow me to present a proposal detailing where I wanted to go and how it would benefit them if they financed my going there.

When I presented my plan to the team at the mine site, I was strongly discouraged. Not only did they refuse to help me, but they implied that I was an idiot to even make this suggestion. Instead of accepting defeat, I refused to give up and called my Head of Safety in the corporate office with all the details. After explaining how opposing forces were fighting against me and disregarding any input I had offered, I asked for their support as I could not afford to remedy the situation on my own.

When I contacted the Ministry of Labor about my work refusal, they suggested a form to fill out in order to get recognition. Knowing that it would take more than thirty days to resolve, this form troubled me greatly. That being said, the Ministry gave me exactly that much time – thirty days - to gather the evidence supporting my claim. This was both a comforting and challenging requirement as I scrambled against time to ensure that I would succeed with all of the necessary steps under their guidelines. In any case, I was determined to find a way and see this through.

With a deep sigh, I made the decision to pursue a study, no matter the cost. I explained to my wife that I didn't care if this meant we lost our house and all of our savings, I needed to do this. Like a light in the darkness, my wife chose to support me wholeheartedly in this difficult endeavor; her confidence and strength encouraged me to push past any doubt. Knowing my partner stood by me gave me the resolve necessary to carry on with this potential endeavor.

After a string of unsuccessful calls to the Ministry of Labor, I decided to contact their higher-ups. My work refusal had reached a point where I felt like I couldn't make any more progress with the Ministry's help. With the Ontario Labor Relations Board being my province's governing body on labor affairs, I called them up in hopes they could help me. Thankfully, they were keen to listen, and soon I got all the support I needed to tackle my work refusal case.

While closely monitoring the events in Ottawa, I noticed a significant shift taking place. It was a bit unsettling to witness these changes unfold. The city's mayor went on television and expressed concerns about escalating racism, mentioning the potential involvement of the National Guard if the situation persisted. The Chief of Ottawa police stated that if any police are caught encouraging these protesters, they could lose their jobs. The mayor's statements seemed exaggerated and lacked credibility. After the press briefings, they claimed that 50% of the police force in Ottawa had left. In reality, some officers had either resigned or taken early retirement. Regardless, it resulted in a considerable loss of personnel for the police force.

Initially, the police and protestors were on good terms. We interacted with them in a friendly manner, exchanging hugs and fist bumps. People even made videos showcasing the cooperation between protestors and the police, where they were helping with various tasks such as cooking, cleaning, and shoveling. However, the situation changed drastically, leading to a significant loss of trust in the police. This was something I personally witnessed during my last visit to Ottawa.

During the final weekend, we encountered numerous road closures, making it incredibly challenging to reach Ottawa. Fortunately, our GPS led us to a secret back entrance, although it extended our travel time from the usual six hours to eight hours. The police were allowing limited access but intentionally making it difficult for anyone to enter or exit the city.

Once we arrived in Ottawa, we noticed a significant transformation. The behavior of the police had changed; they avoided making eye contact and marched through the streets in a military-like fashion, seemingly for show. Meanwhile, the mayor of Ottawa had banned truckers from fueling with jerry cans, which posed a significant challenge during the coldest month of the year. This decision affected the protestors who relied on vehicles to sustain themselves during the demonstrations. Moreover, some provinces and cities had resorted to threatening parents, stating that failure to vaccinate themselves and their children would result in potentially losing custody of their kids. These threats created a tense atmosphere among the people.

In response, individuals began bringing their campers to Ottawa to join the truckers, seeking safety and solidarity within the movement. The community provided ample food and supplies, making monetary transactions unnecessary. However, the mayor imposed restrictions, confiscating jerry cans containing fuel, which resulted in a clever strategy devised by the protestors. Everyone started carrying empty jerry cans, making it impossible for the police to determine which ones were full. It became a tactical move akin to a war strategy. The women even turned empty jerry cans into makeshift purses, carrying their belongings inside. We managed to sneak fuel into Ottawa, assisting in the operation while taking precautions.

However, the situation further escalated when Prime Minister Trudeau declared a state of emergency, addressing various events occurring across Canada as one unified crisis, although they were distinct occurrences. We received information that United Nations planes had landed in North Bay, heightening the tension. As we monitored the events unfolding on our phones, we witnessed a distressing video that went viral on social media.

So, there we were a diverse group of people, united by a common cause, standing shoulder to shoulder with the brave women who had been holding the line against the police. It was a powerful moment, filled with determination and solidarity. As more and more individuals joined the front lines, the atmosphere grew even more charged with energy.

We formed a human barrier, linking arms and refusing to let the police breakthrough. Chants of unity and justice reverberated through the air as we stood firm, resolute in our stance against injustice. It was a testament to the strength of the human spirit, as people from all walks of life came together, putting aside their differences to fight for what they believed in.

The police, armed and equipped in their intimidating gear, tried to push us back, but we held our ground. Their attempts to disperse the crowd with force were met with peaceful resistance and unwavering determination. We knew that our message needed to be heard, and we were not going to be silenced or intimidated.

As the standoff continued, more stories emerged from the crowd. People shared their experiences, their struggles, and their hopes for a better future. We listened to each other, finding solace and inspiration in our collective stories of resilience. It was a moment of human connection amidst the chaos, reminding us that we were all in this together.

Throughout the day, the line held strong, becoming a symbol of our unwavering commitment to change. News of our peaceful resistance spread, capturing the attention of the nation and the world. The bravery of the elderly native women and the unity displayed by the diverse crowd resonated deeply with people everywhere, sparking a sense of solidarity in the face of injustice.

Although the situation remained tense, with sporadic clashes between protesters and the police, our collective spirit remained unyielding. We knew that this was just the beginning of a long journey, and we were prepared to keep fighting for the justice and equality we believed in.

The events that stretched out in Ottawa during those critical days were a turning point in our fight against systemic issues. It was a reminder that real change starts with ordinary people coming together and standing up for what is right. The courage and resilience displayed by the protesters in the face of adversity became a beacon of hope for those seeking a more just and equitable society.

As the days passed, the protests eventually subsided, but the impact they had made was far-reaching. The voices of the people could no longer be ignored, and their demands for change echoed throughout the halls of power. It was a testament to the power of collective action and the unwavering belief that together, we could create a better world.

The events in Ottawa served as a wake-up call for the nation, igniting conversations about social justice, systemic racism, and the role of government in addressing these issues. It was a pivotal moment in our history, one that would continue to shape the path forward as we worked towards a more inclusive and equitable society.

In the aftermath of the protests, the fight for justice and equality continued. The voices that had been raised in unison during those tumultuous days would not be silenced. People remained committed to holding those in power accountable, pushing for meaningful change, and working towards a future where everyone's rights and dignity were protected.

The events in Ottawa would forever be etched in our collective memory as a testament to the power of unity, courage, and the unwavering belief in a better tomorrow. They served as a reminder that the struggle for justice is an ongoing journey and that together, we can create a world where equality and respect reign supreme.

Yet, the quest did not end there. This was just the beginning of me, acquiring the scientific backing I required.

Chapter 8

When It All Came Together

As I meticulously followed the steps of my ambitious plan, I collected masks from work on my designated weekend shift. These masks would serve as crucial samples for the comprehensive bacterial analysis I intended to conduct. To carry out this study, I embarked on a journey to a science lab located in a city about four hours away from my workplace. Over the course of a couple of weekends, I dedicated my time and effort to this project before finally returning home and preparing for my final shift at work.

On that day, having missed an entire week, my primary focus was to collect masks from my colleagues. Throughout the day, I approached individuals in offices and various work areas, politely requesting their masks for the analysis. However, some people hesitated and declined to provide me with their masks. Perplexed by their reluctance, I persisted in understanding their concerns and tried to convince them to trust me and the importance of cooperation. Eventually, they had no choice but to hand over their masks, albeit somewhat reluctantly.

Consequently, I amassed a significant number of masks and proceeded with the analysis. The results were intriguing, revealing that all the individuals who had been in the office had traces of fecal particulates on their masks. These particles likely originated from everyday activities such as touching doorknobs and adjusting their masks, highlighting the need for improved hygiene practices.

Moreover, the analysis uncovered some rather unpleasant discoveries, including the presence of molds. The transition between indoor and outdoor environments, as well as some individuals smoking, likely contributed to the growth of molds. It was astonishing to learn that molds thrive most in a perfect atmosphere characterized by cold temperatures and humid air from our lungs. What's even more surprising is that mold can grow in less than five minutes. To demonstrate the rapid growth of molds, I conducted additional tests with a company, reaffirming the urgency of addressing this issue.

The revelation that everyone's masks harbored mold highlighted the consequences of handling various chemicals throughout the day, leaving our hands constantly contaminated. Even now, observing people wearing masks in public places like the mall, I couldn't help but notice how frequently they touch and readjust their masks, attempting to cover their noses properly whenever they speak. Unfortunately, this behavior inadvertently perpetuated the mold problem. It became apparent that there was much more happening behind the scenes than we initially realized.

Determined to address all my concerns, I also needed to know what we are breathing behind the mask. I contacted an old friend, that I thought could help. This woman, with her responsibility to ensure air cleanliness and maintain records of tests, seemed like the perfect candidate to conduct the oxygen and carbon monoxide test that I desired.

However, she informed me that her equipment was incapable of testing under a mask due to the humidity, which would result in inaccurate readings. Undeterred, she connected me with a senior scientist at a laboratory who possessed the expertise and resources required to assist me. With the senior scientist's guidance, we organized various tests, including monitoring my blood rate while wearing and not wearing a mask during different activities such as running or sitting.

When I mentioned my intention to test oxygen and carbon monoxide levels, the senior scientist assured me that she would find a laboratory capable of conducting these specific tests. It took a few days of diligent effort. Additionally, I must note that the process incurred expenses, as outlined in the Green Book guidelines. To navigate this aspect, I collaborated with my joint health and safety committee, presenting them with a well-thought-out plan. However, to my disappointment, my worker rep and co-chair made it clear, that the company representatives said I was not welcome in any meetings, further hindering my progress. My co-chair told me I could present something to our committee, but the next meeting wouldn't be for weeks, leaving me late for my deadline. As well I asked to use the computers at work, and my co-chair told me in a text that management said I was forbidden to use any computers.

Growing increasingly frustrated, I decided to contact the ministry and report that I had exercised my right to refuse work and the steps were not being followed. The ministry directed me to an online form, but I harbored doubts about receiving substantial assistance from individuals who were occupied with their own protocols, such as taping masks to their faces. Moreover, these endeavors were beginning to impose financial burdens on me.

Seeking a resolution, I brought up the issue with the mine, emphasizing that they should bear the expenses incurred during this process. Unfortunately, their response was far from supportive. They bluntly refused to assist me and instead told me to handle the matter independently. To add insult to injury, I received messages from my friends in high positions who had previously supported me, expressing their inability to continue offering assistance. They divulged that a meeting had taken place behind my back, during which everyone involved was threatened with termination if they dared to aid me any further.

So, I called the deputy doc again, showing him the text messages as proof that there were individuals conspiring against me. It became evident in my mind that he was worried about being exposed if I succeeded in my endeavors. It seemed like he didn't want me to be successful either, as he kept sending me in circles, providing no real solutions. At some point, I reached a breaking point and decided to ignore everyone's interference. I confided in my wife, telling her that I didn't care anymore. I was willing to pay for the study myself and prove them wrong, without relying on their help.

Taking matters into my own hands, I decided to call the bosses at the Ministry of Labor. I took a step further and contacted the Ontario Labor Relations Board, which is the governing body in my province responsible for ensuring proper adherence to regulations. It functions like a courtroom, with judges presiding over cases related to labor disputes and violations. When I applied for a hearing, the judge assigned to my case expressed genuine interest in my story and decided to grant me a hearing.

It happened right before the expiration of the 30-day period, while I was still at work. I received a call to go up from underground as the ministry wanted to see me. However, I decided to make them wait all day, as they kept putting me on hold for hours, playing games with me. I played games with them in return, asserting that my work took priority over their visit. The defiance was clear when I openly declared over the radio that the ministry could wait because I had important tasks to attend to.

Eventually, I made my way to the meeting room where the ministry representatives, managers from the company, and my workplace were gathered. Prior to this encounter, I had undergone retraining on Covid protocols, and during that process, I had learned about the importance of social distancing, despite the inconsistencies in their policies. As I entered the room, I asked the manager if the Covid policy had changed in the last month, to which they replied negatively. With that confirmation, I took off my mask and threw it onto the table before taking a seat and propping up my feet. The unexpected action caught everyone in the room off guard, as they were not accustomed to such defiance.

The ministry representative was visibly displeased with my actions, with wide-open eyes displaying his surprise. Nonetheless, I confidently addressed the room, challenging them to discuss the matter at hand. The ministry representative attempted to reprimand me for removing my mask, but I stood my ground and reminded him that he had refused to listen to my evidence earlier. I declared that I had already filed a section 52 application and would be serving him and the company with the necessary legal documents the following week. I assured him that the court case was already scheduled and they would be receiving notice soon. With that, I dismissed their attempts to engage further, requesting permission to return to work and refusing to listen to any more of their excuses.

Feeling overwhelmed with anger and frustration, I expressed my discontent to my boss, who was also my friend. I declared that I had reached my limit and had no intention of returning to work. I decided to go home instead, needing some time to calm down. My boss understood my state of mind and allowed me to leave, saying, "Okay, go home, Rylen."

As I made my way out of the mine, one of the safety representatives approached me, astonished by what had transpired during the meeting. He informed me that they were furious and had warned him that if I ever acted like that again, I would be kicked off the property. Curious about who had made such threats, I asked him to point out the person responsible. He gestured toward one of the individuals present in the meeting room, indicating that he was the one who had issued the warning.

Determined to confront the individual directly, I walked up to him and inquired if he had a problem with me. He confirmed his disdain, questioning my motives and actions. I responded by explaining that I intended to introduce legislation to protect people from being forced into similar situations in the future. Furthermore, I emphasized that I was fighting for my children's rights and vehemently opposed seeing masks on them, considering it as a form of child abuse. Frustrated and upset, I left that day with a heavy heart.

During that time, there was a peculiar week when a crew member contracted Covid-19. The company had a tracking system in place to identify individuals who may have come into contact with the affected person. However, due to an oversight, my name was not included on the list. One day, we were all called out from underground and instructed to proceed to the office. Upon reaching the surface, I was surprised to see a setup of plastic pathways and personnel in large biohazard suits, resembling something out of a science fiction movie.

Already frustrated by the company's misguided measures, I couldn't help but express my annoyance. I addressed the rest of the crew, stating that if we were given even a single day off due to this situation, I would quit and be done with the company. As we reached the surface, the safety manager informed us that someone on our crew had tested positive for Covid-19 and that some of us would need to go home for two weeks to quarantine and get tested.

Furious with the news, I erupted in the hallway, adamantly refusing to undergo any testing. I made it clear that I would not allow them to put any tests near my nose. I declared that I was sick of their nonsense and dramatically quit by throwing my light at the safety manager before storming off toward the exit. However, the gentleman called out my name, informing me that I was not on the list of individuals requiring quarantine and testing.

Despite the chaotic turn of events, my determination to fight for what I believed in remained steadfast. The upcoming court hearing promised to be a crucial turning point in my battle for justice and the rights of my children.

Chapter 9

A Beginning to the End

In the face of continuous obstacles and blatant disregard for my rights, I made the decision to take matters into my own hands. I notified the ministry of my completed study, highlighting the inaction and denial I faced from the Joint Health and Safety Committee (JHSC) in presenting my findings. The ministry's questionable advice prompted me to bypass their authority altogether. I took a leap of faith and reported both the mining company and the ministry's negligence in adhering to the laws outlined in the green book. I sought a hearing with the Ontario Labor Relations Board, eager to present my conclusions and advocate for justice.

The road to the hearing was arduous and demanding, requiring countless hours of dedication and personal financial investment. Amidst my preparations, a peculiar turn of events occurred—COVID-19 seemed to vanish into thin air. Masks, once mandated, now became a mere recommendation. It was as if the very fabric of my quest was intertwined with the threads of this global pandemic.

The stage was set for the moment I had been tirelessly working towards.

Finally, the day of the hearing arrived, shrouded in confidentiality. While I cannot disclose the specific details of the proceedings, what I can share is that the court ordered the mining company to allow me to tell my story to the mine. This victory, albeit partial, fueled my spirit and reaffirmed the importance of my mission.

Essentially, at the beginning of the court session, they explicitly prohibited any recording. We agreed to comply. Then they proceeded to explain that the purpose of the hearing was to attempt mediation. I expressed my desire to receive an answer to a question I had posed a year ago, which led to this case. I was instructed to provide evidence to validate my concerns. To support my case, I possess recordings, emails, and other evidence indicating that the government never conducted a study on the matter. During the court proceedings, I reiterated my request for an answer to my question, but the response I received was that the COVID situation was over, so they did not have to address it.

In essence, I pointed out to the lawyers that their response essentially answered my own question, as it revealed that they couldn't find an answer because the study I sought never existed. Furthermore, when discussing potential changes, I expressed my desire to modify legislation. However, I was informed that it was no longer necessary or possible. In response, I requested permission to share my story with the world, and the judge granted me that privilege. Therefore, in addition to addressing the implications of mask usage, the judge authorized me to inform minors about these consequences.

Throughout this journey, I dedicated countless hours and invested substantial financial resources to acquire answers from Ontario's top doctors and scientists. Their recognition of the significance of my study filled me with a profound sense of pride. I became the pioneer, the bearer of truth in a world where deception had cast a shadow over the populace. The government had coerced the Canadian people into wearing masks without any scientific validation to support their claims.

As I conclude this chapter of my story, I feel compelled to deliver a message to all who may read these words. Be brave, stand up against bullies, and utilize the strength found in the bonds of brotherhood. Let us invoke the spirit of accountability and demand transparency. Each and every one of us possesses a voice that must be heard. In my role as the JHSC representative for my crew, I vow to amplify the voices of my fellow workers and advocate for their rights and well-being.

I remain hopeful that, with the proper support from the JHSC and the Ministry of Labor, we could have exposed this giant lie and protected not only the site I worked at but the entirety of Canada. My journey has shown me the power of grassroots efforts, the significance of unwavering dedication, and the potential for change when individuals rise up to challenge the prevailing norms. It is with this newfound purpose that I eagerly anticipate continuing my role in the JHSC and fighting as fiercely as I can to help fix the system and safeguard the rights and well-being of all workers.

In unveiling the truth, we ignite the spark of change. Let our voices reverberate, let our actions resonate, and together, let us forge a better future.

As I reflect upon the extraordinary journey I embarked upon, a myriad of emotions floods my senses—determination, frustration, and finally, a resolute sense of accomplishment. This is the story of my relentless pursuit of truth, fueled by curiosity and an unwavering belief in the power of standing up for what is right. It is a tale that unveils the hidden realities surrounding public health and highlights the potential for grassroots efforts to effect change on a grand scale.

Now, as I pen the final words of this chapter, I do so with a heart brimming with gratitude for the journey I have undertaken. The twists and turns, the triumphs and tribulations, have all woven together to create a tapestry of growth and self-realization. And in this exquisite culmination, I stand upon the precipice of a future brimming with limitless possibilities.

For it is through the genesis of my journey, that serendipitous moment when everything clicked into place, that I have not only discovered my truest self but also the profound interconnectedness that binds us all.

Ladies and gentlemen,

As we come to the end of this remarkable journey, I stand before you to share my final thoughts, and my closing remarks on the pages we have turned together. This story, my story, is one of determination, resilience, and the unwavering pursuit of justice.

Throughout these chapters, you have accompanied me on a voyage where I dared to challenge the norms, question the narratives, and stand up for what I believed in. It was never an easy path. The road was paved with obstacles, resistance, and even personal attacks, but I never allowed myself to falter.

In a world where science should be the guiding light, and where curiosity should be celebrated, I found myself in a place where asking questions became a dangerous act. The very foundations of scientific inquiry were shaken as assumptions and predictions were presented as irrefutable truths. It was a disheartening realization, but I knew I couldn't remain silent.

I delved deep into the research, reaching out to esteemed laboratories and organizations in search of answers. The studies I uncovered shed light on the potential dangers of the very masks we were told to wear without question. It was a revelation that stirred within me a sense of responsibility, a duty to my coworkers and to all those whose voices were silenced.

I confronted the system, demanding transparency, demanding the science that should have guided the decisions made on our behalf. But my pleas fell on deaf ears, met with resistance and threats. Yet, I refused to be intimidated. I sought refuge in the Green Book, the invaluable guide that outlines our rights and the steps to ensure our safety.

When the authorities turned a blind eye, I took matters into my own hands. I embarked on a quest for truth, seeking out experts and scientists who could help validate my concerns. With every conversation, and every recorded response, I gathered the evidence that would expose the deception and unveil the truth.

But the journey was far from over. As I fought for justice, I encountered roadblocks and bureaucracy. The very institutions that were meant to protect us seemed complicit in the silence. Yet, undeterred, I persisted, navigating through the maze of red tape, determined to bring my findings to light.

And so, here we stand, at the end of this gripping tale. My study, my pursuit of truth, has uncovered a startling reality. The very authorities who claimed to have scientific evidence to back their mandates had no such evidence to offer. It was a revelation that shook the foundations of trust.

I share my story with you today not as a call to rebellion, but as a reminder that our voices matter. We must dare to ask questions, to challenge the status quo when it fails to align with what we know to be true. Our rights, our well-being, and the safety of our fellow workers depend on our courage to speak up.

In closing, let us remember that knowledge is power, and our responsibility lies not just in acquiring that knowledge but in sharing it. May this story inspire us to seek the truth, question authority when necessary, and forge a future where transparency, safety, and justice prevail.

Thank you for being a part of this journey.

Made in the USA
Monee, IL
19 September 2023

42974966R00044